MILK
DONALD CARRICK

Greenwillow Books, New York

FOR THE McCOYS

Printed in Hong Kong by
South China Printing Co.
First Edition
10 9 8 7 6 5 4 3 2

Library of Congress Cataloging in Publication Data
Carrick, Donald.   Milk.
Summary: Describes the entire process by which
milk goes from the cow to our homes.
1. Milk—Juvenile literature. 2. Dairying—Juvenile
literature. [1. Milk. 2. Dairying] I. Title.
SF239.5.C37   1985   637.1   84-25879
ISBN 0-688-04822-6   ISBN 0-688-04823-4 (lib. bdg.)

In the summer the cows live outside in the pasture. They spend most of their time eating. Winters, the cows live and eat in the barn except to exercise in the yard each day.

When the day ends, the cows start moving toward the barn. At the road, the farmer opens the gate. All traffic waits until the cows are safely across.

In the barn each cow finds a stall. When they are all in, the cows are fed grain and their udders are washed. While they are busy eating, the farmer attaches a milking machine to each cow.

The milk from each cow is poured into a transfer machine from which it is pumped into a cool holding tank.

After the evening milking the cows go back to their pasture. They eat grass and sleep until morning.

At dawn the farmer gets up. The cows
are waiting to be milked again.
It is still early when the tank truck
comes to collect the milk. Each day
the truck takes the milk to the dairy.

At the dairy the milk is pumped to the pasteurizing machine which quickly heats the milk to kill germs. After cooling, the milk is mixed in the homogenizing machine to keep the cream from separating.

The next machine forms cartons and fills them with milk. At the same time that it closes the cartons, it stamps them with the date.

Drivers load crates
of the fresh milk into
refrigerated trucks.

The trucks deliver the milk
to the stores. It is ready
for everyone to drink.

In the store the milk is kept cool
in the dairy case till customers
come to take it home.